Study Guide for

Alive and Well at the End of the Day

The Supervisor's Guide to Managing Safety in Operations

PAUL D. BALMERT

Published by John Wiley & Sons, Inc., Hoboken, New Jersey.
Published simultaneously in Canada.

For general information on our other products and services please contact our Customer Care Department within the United States at (800) 762-2974, outside the United States at (317) 572-3993 or fax (317) 572-4002.

Wiley also publishes its books in a variety of electronic formats. Some content that appears in print, however, may not be available in electronic formats. For more information about Wiley products, visit our web site at www.wiley.com.

Library of Congress Cataloging-in-Publication Data:

Balmert, Paul D., 1950–
 Alive and well at the end of the day : the supervisor's guide to managing safety in operations / Paul D. Balmert
   p. cm.
  Includes index.
 ISBN 978-0-470-46707-7 (cloth)
 1. Industrial safety—Management. 2. Work environment—Safety measures. I. Title.
 T55.B254 2010
  658.4'08—dc22   2009037095

10 9 8 7 6 5 4 3 2

Study Guide

Alive and Well at the End of the Day

"Experience ***is*** *the best teacher."*

--Anonymous

Alive and Well was designed to serve as a practical guide to safety leadership: its focus is not on explaining leadership *theory* but on teaching leadership *practices*, the "what to do" and "how to do it." In the book, safety leadership practices from starting off the day with a safety meeting to adding up the score at the end of the day are taught in a series of chapters, each one its own, self-contained lesson.

That's how the book teaches the lessons. But life as a leader doesn't come at you like the chapters in a book. So, a more important question for you to ask would be "What's the best way for me to learn, understand, and master the practice of safety leadership?"

Teaching and learning may be related, but the processes to accomplish each are entirely different. How do you learn something? More importantly, how do you come to understand something? Most importantly, how do you become really good at something?

Undoubtedly you know the answers.

You might read about something in a book, or hear it explained, but you won't really understand it until you try doing it yourself. That's true for everything from what you learned as a kid, such as tying your shoelaces or throwing a football in a perfect spiral, to what you've learned as a leader in business, like leading a safety meeting or intervening when someone is taking too much risk. Think about what actually happens when you consciously determine to learn how to do something new. As you approach the point of doing something new or different, you essentially teach yourself. That teaching takes place inside your head, normally in the form of a conversation. You first say to yourself, "OK, let's try doing it differently. First, I'm supposed to… Next, what I do is …"

Then you actually *do* it for the first time. And then you get feedback.

It might work perfectly. You say to yourself, "That was easy." Or it might fail miserably. You think "I'll never try that again." Most of the time, first-time experience falls somewhere in between those extremes. It's different, but far from perfect. You say, "OK, so the next time, I'm going to…" Then, you do it again, a little bit differently and a little bit better. You might even get pretty darn good at doing it.

Experience really is the best teacher. We come away from an experience understanding what we know and appreciating what we don't fully understand or can't effectively perform. That is, If and it's often a really big "if"—we take the time to think about what we have learned. And what we haven't learned. All too often, we're just too busy to take the time to do that.

So, learning becomes a hit or miss proposition. We might get lucky, subconsciously getting better at something without really trying. Often, we wind up not learning anything new. That leaves us relying on what we already know. Great if you're satisfied with the present state, but not good if you want to get better. As Einstein put it, "Insanity is doing the same thing over and over again, and expecting different results."

Safety leadership is no different. If you want to get better safety results, you have to lead better. If you want to lead better, you have to first learn f, practice second, and evaluate third. Then do it all over again. It is in that process that *understanding* and *mastery* are attained. .

Do you want to lead better? If so you need to master the practice of leadership.

In the case of managing safety performance, these practices are described in *Alive And Well At The End Of The Day*. Reading and understanding them initiates the learning process. But it will help you get the maximum out of reading the book to realize that the real understanding and mastery of the subject takes place at the point where the tools and concepts are put in play: out in the real world and real time of the organization you lead.

This Study Guide is designed to help you master the tools and concepts found in the book as you put them in play. It provides a set of discussion questions that you can ask yourself *before* and *after* you've applied the tools and concepts. The questions will help you succeed when you put the tools to use, and better learn from the experience of using them. The questions have been divided into two sections: **Preparation Questions**, designed to prepare you to put the tools in play, and **Debrief Questions**, which help you focus on learning from your experience using the tools.

The Study Guide also lays out a structure and process by which you can engage others who have the same interests as you in mastering the tools. Yes, you can follow this Guide on your own, but there's power in numbers. Sharing common experience with others—explaining what you are learning, and listening to their experiences—is guaranteed to enhance your learning.

The Study Guide follows the order of chapters in the book, and provides the kind of structure and process to move from *knowing* something to *understanding* it to being able to *execute* it effectively. The chapters in the Guide need not be followed in any particular order though. This enables a user to follow a chapter at the time when there is the need. Did you roll out a new safety policy this week? Find the chapter on Communicating Safety Policies. Are you in charge of the department safety meeting this month? Read the chapter on Safety Meetings, and see what tools you might use to make the most of the time.

Chapter 1

A Guide To The Guide: Getting The Most Out Of This Book

No matter where in the world they work, or what kind of business they're in, leaders in operations face the same fundamental set of tough safety challenges. Whether it's getting everyone to follow all the rules all the time, avoiding complacency, getting everyone to report near-misses, or finding the time to actually *manage* safety performance, these challenges are brutally tough – in large part because they're relentless – and timeless.

The good news is that the most successful leaders have figured out how to deal with them, and doing that is largely a matter of practice...leadership practice.

Preparation Questions

1. How does the list of "tough safety challenges" on page 2 match up with those you face as a leader in seeing to it that everyone goes home alive and well at the end of the day?

2. What are the factors that make these challenges tough for leaders in operations?

3. What are the specific challenges you are looking for help with?

4. How will you be able to determine how successful you are in dealing with these challenges?

You will find the **Debrief Questions** in the final chapter of the Study Guide.

Chapter 2

The Case For Safety

The Case For Safety answers the simple – but critical – question every leader faces: Why is safety your most important goal in operations? It's a question that every leader knows the answer to, but not every leader understands. If they did, the world of work – and the practice of safety – would look very different.

The answer is found in three simple questions:

What are the most important things in your life?

How would those things be affected were you to suffer a life-changing accident?

Is anyone working for you whose answers would be different from yours?

Once you understand these, your mandate becomes clear: Send everyone home safe at the end of every day!

Preparation Questions

1. What are your answers to the first two of those simple questions?

    What are the most important things in your life?

    How would those things be affected were you to suffer a life changing accident?

2. Who do you know who has suffered a serious accident? What kind of impact did it have on that person and his or her family?

3. Why do you think so many leaders in operations fail to fully understand *The Case For Safety* until someone gets seriously injured?

4. When you ask those same two simple questions to the members of your crew, how do you think they will answer?

Debrief Questions

1. Who did you pose the questions to?

2. How did their answers compare to what you were expecting to hear?

3. What impact did asking those questions have on their understanding about what makes safety important?

Chapter 3

The Practice Of Leadership

What leaders do to lead other in all walks of life boils down to "use words" and "take action." Every follower on the planet knows that "actions speak louder than words." That's because actions are things that followers can actually see a leader do — like following the rules and fixing problems. Using words — language — to communicate is a far more complex leadership process: not as important as taking action, but still very important. Words are a leader's way of explaining, eliciting, exciting and engaging followers.

One very important leadership tool that uses words is the "Stump Speech." A Stump Speech is a concise statement of a leader's beliefs, values, advice, and expectations about working safely. It need not be long, but should have enough meaningful content to make it more than just a slogan.

Preparation Questions

In order to develop your Stump Speech for safety, you need to give some thought to its content — what is the message to your followers about working safely? Here are four questions that can help you get your arms around your message:

1. What are the most important reasons the members of your team should work safely?

2. What are the most important things the members of your team should consider or do in order to work safely?

3. What is your best advice about how to work safely?

4. What are the important things you will do as the leader to make sure no one gets hurt on the job?

A Stump Speech has three simple parts:

1.   Opening

2.   Content: Your key points

3.   Closing

Debrief Questions

1. When did you give your Stump Speech?

2. What kind of reaction(s) did you get?

3. What changes – if any – should you consider making? What's missing? Needs clarification? A better explanation?

4. When might be the next good opportunity to put your Stump Speech to effective use?

Chapter 4

Moments of High Influence

The concept known as Moments of High Influence is simple – and powerful: "In the everyday life of a leader, there are times and places where he or she is in a natural position of leadership and followers are in a high state of readiness to be influenced." As a leader, when you recognize these moments for what they are—ready-made leadership opportunities—you're bound to raise your game. The secret is to recognize Moments of High Influence in the moment they occur...rather than in hindsight. Doing that requires a leader to be able to put himself in the follower's shoes, seeing the situation from the follower's viewpoint.

Sounds simple. But it is easier said than done!

Preparation Questions

1. Every leader is some other leader's follower. It's easy for a leader to recognize the Moments of High Influence his or her leader has. Why do you think it is it often difficult for leaders to recognize their own Moments of High Influence?

2. What safety-related Moments of High Influence have you personally experienced as a follower that have made a big impression on you?

3. What was a recent Moment of High Influence you had as a leader?

4. What specific Moments of High Influence do you think you might experience in the next few days?

5. Knowing that each is a Moment of High Influence, what might you do differently than you normally would?

Debrief Questions

1. What was one recent safety-related Moment of High Influence?

2. Recognizing it was a Moment of High Influence, what did you do differently compared to how you would have normally handled the situation?

3. How well did that work?

4. What have you learned from the experience?

5. What else might you do—or do differently—the next time a situation like this comes up?

Chapter 5

Managing By Walking Around

Safety performance always gets better when the supervisor is out on the job seeing what's going on, and just as important, being seen by those doing the work. The challenge is finding the time to actually do that, as supervisors are some of the busiest people on the planet.

That's where Managing By Walking Around fits in: it's the calculated use of the time and presence of the leader. For safety, one of the best ways to perform the procedure is to consider the most likely reasons why people get hurt doing the kind of work your team performs. Once you understand what those most likely reasons are, the best places, times, situations and people to observe become rather obvious.

Then the game shifts to observation: analyzing the details. What are you seeing that's good...and bad?

Preparation Questions

1. What are the top ten reasons why people get hurt doing the kind of work your team performs?

> 1.
>
> 2
>
> 3.
>
> 4.
>
> 5.
>
> 6.
>
> 7.
>
> 8.
>
> 9.
>
> 10.

2. To be able to see what real performance looks like for each of these questions, and based on what you know about the work your team performs, how would you structure your Managing By Walking Around efforts (think when, where and who)?

    1.

    2.

    3.

    4.

    5.

    6.

    7.

    8.

    9.

    10.

Debrief Questions

1. Where did you go in your Managing By Walking Around?

2. Why did you choose to go there?

3. What did you see when you were there that was good?

4. What did you see when you were there that was not good?

5. Based on what you saw and learned, where else should you go?

Chapter 6

Following All The Rules All The Time

If you trace every safety rule ever written back to its point of origin, you'll invariably find a tragedy, large or small. In that sense, safety rules are different from all of the other organization's policies and procedures. But that difference doesn't make compliance—getting everyone to follow all the rules all the time—easy. A complicating factor is that there are four separate requirements that need to be met in order for people to comply with the rules.

1. Those expected to follow the rules must first *understand* them.

2. They must be able to *remember* the rules.

3. They must recognize that   a rule applies to the situation they find themselves in.

4. They must choose to *comply* with the rule.

Each requirement poses its own set of unique challenges.

Preparation Questions

1. What specific safety rules do you find your team has the most difficulty following?

2. What are the reasons why compliance with these specific rules is particularly difficult?

3. How would you rate the effectiveness of the training your people receive for these particular rules?

4.   What actions can you take as a leader to improve compliance with the safety rules?

5.   What's your Stump Speech for following all the rules all the time?

6.   How can you determine if compliance is improving?

Debrief Questions

1. What specific actions did you take to improve compliance with the safety rules?

2. How well did you execute those actions?

3. What impact have they had on compliance?

4. What else might you consider doing to further improve compliance?

Chapter 7

Recognizing Hazards And Managing Risk

While the terms hazard and risk are sometimes used interchangeably, there is a vitally important difference between them. A *hazard* is a source of danger; *risk* is the probability that the hazard will cause an event. In the case of safety, the consequences of an event range from a very minor near-miss to a fatality.

Safety rules are built around known hazards. But following the rules doesn't guarantee safety, as there are plenty of ways to get hurt that fall outside the safety rules. A supervisor needs to understand what creates hazards, the limits of protection afforded by the procedures, and how to recognize the hazards most likely to send someone home hurt or hurting at the end of the day.

Preparation Questions

1. Most people who get hurt were very knowledgeable about the hazard that caused their injury. Why do you think that is so?

2. Why do you think it is difficult for people to perceive the hazards most likely to send them home hurt – or hurting?

3. In your operation, what do you think are the hazards most likely to wind up sending someone home hurt?

4. What are the key hazard recognition and management procedures or processes that you and the members of your team regularly use and rely on to identify job hazards?

5. Applying TIE—the three routes of injury—to those policies, how good a job do they do in identifying and managing the hazards capable of getting someone hurt?

6. What actions can you take as a supervisor to increase the ability of the members of your crew to recognize hazards and perceive risk – for the tasks they perform?

Debrief Questions

1. What specific actions did you take to better understand how well your hazard and risk management processes are working out on the job site?

2. What did you learn in the process?

Chapter 8

Behavior, Consequences – And Attitude!

As a leader, when you see someone working unsafely—either by not following the rules or taking too much risk—the right thing to do it to intervene. A successful intervention not only will get the person out of harm's way, but can alter the person's future choices and behavior. The intervention strategy summarized in a simple, five step conversation is the approach some of the most successful leaders have followed to accomplish both goals.

Often overlooked in the process of managing behavior is the power of positive feedback from the leader. With one small modification, that same strategy can be employed to give positive feedback for good behavior. Instead of asking why the person is not working safely, find something specific about the person and the situation that is positive, so as to reinforce safe behavior. That makes the compliment genuine.

The majority of leaders think the best way to get people to stop working unsafely and start working more safely is to attack what they see as the root of the problem: attitude. It's a great theory, but one that is incredibly difficult to successfully put into practice. If attitude is simply defined as "what someone thinks," it's nearly impossible to know for certain what's going on in someone's mind. And at, best, it's extremely difficult to change someone else's mind! The better strategy is to focus your effort on what you can see, hear, and manage: behavior.

Finally, never lose sight of the fact that consequences do matter, and the consequences that matter most are found in the Case for Safety: going home alive and well at the end of every day.

Preparation Questions

1. Why do you think changing attitude is such an appealing "target" for so many leaders?

2. What systems and processes (performance appraisals, attitude surveys, for example) does your organization use that focus attention on "employee attitude?"

3. If you are struggling with getting compliance on a particular safety policy or procedure, ask yourself this question: What are the *positive* consequences for *not* following the rule?

Debrief Questions

1. When did you provide safety behavior feedback to someone in your organization? Was it positive or corrective?

2. How closely did you follow the five steps of the intervention strategy?

3. What steps did you find easy to follow? What steps did you find more difficult to follow?

4. The next time you intervene, what will you do better or differently?

Chapter 9

The Power Of Good Questions

Leading by means of asking questions is a great way to gain leadership leverage – increasing your influence as the leader without putting more effort into the process. The way to take advantage of the power found in asking questions is to ask Darn Good Questions.

A Darn Good Question has three elements. The first is purpose: the primary objective of A Darn Good Question is not to get information, but rather to lead and influence a follower with some specific end in mind. The second element goes hand in glove with purpose: a specific audience targeted by the purpose of the question.

The third element is the question itself: A Darn Good Question is based on one of the key words: who, what, when, where, how or why. Each one of these key words in a question assures an answer other than a simple yes or no. Moreover, each of these words will send the target audience in a specific direction as chosen by the leader.

Preparation Questions

1. Why do you think most leaders are more inclined to tell their followers what they think, rather than lead by asking Darn Good Questions?

2. When has a leader you worked for asked you what you now recognize as a Darn Good Question? What impact did that have on you?

3. What are some Darn Good Questions you can ask some of your followers?

Purpose	Audience	Darn Good Question

Debrief Questions

1. What situations did you choose as your opportunity to lead by asking Darn Good Questions?

2. What specific questions did you ask?

3. How well did each specific question work in achieving your purpose?

4. Based on your experience, what have you learned about the process of asking Darn Good Questions?

Chapter 10

Making Change Happen

Every time a supervisor stands up front of the crew, announcing that, "We have a new safety policy," it represents a Moment of High Influence. But communicating change will almost always generate some amount of resistance as the crew doesn't have any choice but to go along with the change. People don't resist change...they resist *being* changed.

Assuming the change affects the crew, the supervisor manages the most important step in the change process: execution, making the change happen. Whenever a policy is changed, the supervisor's goal is to have every team member understand what the change is, and be committed to making the change.

Understanding that, there are two simple rules to follow that will make the change process go better. The first is to tell people why the change is being made. Explaining why reduces resistance as people are far more likely to go along with a change when they know the reason. In the case of safety policies and procedures, the reason for the change almost always has something to do with an accident. The second rule is to focus on the question "What do we need to do to make the new policy work?" That's what making change happen is all about.

Preparation Questions

The perfect time to apply this chapter is at the point where you will be communicating a new or revised safety policy or procedure. When you're called upon to do that, consider the following questions before you communicate the change.

1. What caused the policy to be written or changed?

2. What issues or questions can you expect to be raised by those who are impacted by the change?

3. How will you begin your communication session?

4. How will you close out the session and assure understanding and commitment?

Debrief Questions

1. How well did you follow your outline for communicating the change?

2. What steps in the process went well?

3. What problems did you encounter in explaining the change, getting buy-in, or following your communication plan?

4. What have you learned from the process that will help you the next time you communicate a change?

Chapter 11

Managing Accountability

Every leader knows "holding people accountable" is an important part of managing safety performance. Successfully managing accountability begins by understanding what the term really means. Then it requires understanding the steps in the process to "hold someone accountable."

Managing accountability is done on the basis of cause and effect. When something happens, it's normally because someone did something (or didn't do something they were supposed to do) to cause it to happen. And, when something happens, there are always effects. "Responsible" ties back to the cause. "Consequences" are found in the effects, and often describe corrective action." "Accountable" simply means answerable: requiring someone to give an account.

Holding someone accountable is about asking the questions that need to be asked. Not just about what happened, but *why* it happened, how someone could allow something like that to happen, and what they have learned from the experience. Accountability is normally brought up in the aftermath of failure, but it doesn't have to be that way. After a success, it may be appropriate to attribute accountability by recognizing positive performance. The odds are good there are human fingerprints on these causes, too.

The Five S's provide the structure for your discussion to manage accountability. The first three S's__ situation, significance and specifics—set the stage for the core of the process: asking the questions that need to be asked. The natural place to start is by asking the person for "his side of the story," and then, of course, digging deeper by asking Darn Good Questions. Finally, holding someone accountable must include identifying the steps to prevent the problem from happening again, or sharing the learning from a positive event. That's the final step in the process.

Preparation Questions

1. "Managing accountability" is often confused with the administration of corrective action or discipline. They are related, so it's important to understand the latter as well. In your organization, what are the *potential* consequences for not following the safety rules or taking too much risk?

2. Thinking back over your experience, when have you attempted to actually "hold someone accountable," as opposed to "administering consequences" in common forms such as counseling, oral or written warnings and even time off?

3. How well did that approach work? How closely did it follow each of the Five S's?

4. Looking forward, what specific situations do you foresee which lend themselves to holding someone accountable?

5. If you follow the Five S's, what will you say, –or ask, at each step in the process?

Debrief Questions

1. What was the specific situation for which you followed the Five S's?

2. How well did you follow each of the steps in the process?

3. As you followed the process, what steps proved easy? What steps were more challenging?

4. What were the important questions you asked in the forth step of the process?

5. The next time you hold someone accountable, what can you do even better?

Chapter 12

Safety Meetings That Matter

A safety meeting—whether it's the morning tool box safety meeting, the monthly department safety meeting, or the annual "all hands" meeting—is in theory a Moment of High Influence. But, if it's not a good safety meeting, what kind of influence will it have out on the job?

One sure way to hold an effective safety meeting is to get your audience to do the talking for you. You want them talking about what *you* want them to—not what *they* would like to—and, in the process, actually accomplish something important.

Ask, Don't Tell is a simple and effective way to do exactly that.

Like asking a Darn Good Question, this approach requires that you are clear as to exactly what you want to accomplish by talking about the subject. The next step is to tell the long story short, keeping you out of the presentation business. Usually the details are unnecessary to your purpose, so it should take but a few moments to set up the subject with a headline and summary.

Before asking Darn Good Questions, make the connection by telling everyone why they should pay attention to the topic and participate in the discussion. Finally, two or three Darn Good Questions are normally sufficient to cause fifteen minutes of good discussion on the subject.

Preparation Questions

Pick out a topic for an Ask Don't Tell Safety Meeting, and then complete the outline.

Purpose: *My goal in getting my team to talk about this subject is so that...*

Headline:

Summary:

Connection: *"The reason why it is important for us to talk about this topic today is because….*

Darn Good Question 1:

Darn Good Question 2:

Darn Good Question 3:

Debrief Questions

After you've tried conducting a safety meeting using the Ask Don't Tell method, answer these questions:

1. What topic did you choose?

2. When and where did you conduct the safety meeting?

3. In the meeting, how long did each of the sections (headline, summary, connection and questions) take?

4.   How closely did the meeting follow your Ask Don't Tell plan?

5.   What worked well?

6.   What Darn Good Questions did you ask? On a scale of 1 to 5, how well did each one work?

7.   The next time you use this approach, what will you do differently?

Chapter 13

Managing Safety Suggestions

When a supervisor gets a safety suggestion, it's a Moment of High Influence…for the employee who makes the suggestion. For the leader, it's another opportunity to lead. It also means dealing with a problem, because a safety suggestion is the combination of a problem and a solution.

The first consideration in dealing with any safety suggestion is that of priority or urgency: how quickly you need to deal with the suggestion. That determination should be based on the nature and severity of the problem. You always want to say "thank you" (not "good idea") to the person who has submitted the suggestion: that's how you recognize the Moment of High Influence.

As to what to do about the suggestion, there are always plenty of options, ranging from "let's do it," turning in a formal suggestion, or letting the employee champion the idea. It's always a good idea to look at the situation giving rise to the problem. That's Managing By Walking Around. If the solution finds its way into practice, be sure to bring it to the attention of others. That's a great way to encourage others to turn in suggestions.

Preparation Questions

1. How often to the members of your crew turn in safety suggestions?

2. How a safety suggestion is normally handled in your operation?

3. How well does that work?

4. What kind of safety problems would you like to see brought up by suggestion(s) from your crew?

5.   The next time you receive a safety suggestion, what will be your plan to respond?

Debrief Questions

1.   What was one specific safety suggestion you received?

2.   What was the problem? The proposed solution – if you received it?

3.   What specifically did you do when responding to the suggestion?

4.   What went well?

5.   What might you do differently the next time you receive a safety suggestion?

Chapter 14

Creating The Culture You Want

Every operation has a safety culture: "the way things really are around here." If you want to know what kind of safety culture you have, all you have to do is look—not at what any one person does, but at what "everybody does." It isn't necessary to look at everything. Pick out a few key dimensions, like people stopping at the stop signs, carefully executing job safety analyses, and reporting near-misses... and look. What you see is your safety culture.

If what you see is not a "culture of safety," you need to be in the culture *change* business. Left unattended, a culture will slowly change but in a direction of it's choosing. You want *transformation,* which implies both speed and a change in a direction of your choosing.

The first step in transforming culture is to be very clear about the change you want. Define your desired end state in terms of specific and observable behavior, not as values and beliefs (which are highly resistant to being changed).

Then, build your plan. Identify the leverage points that make the process more likely to succeed: key leaders working down in the chain of command; key events in the everyday life of an organization where people are in a high state of readiness to be influenced—Moments of High Influence! Sometimes small but significant problems or processes, which if changed can create a powerful ripple effect through an organization. Don't overlook your power to influence your organization through leading by example.

Preparation Questions

1. In terms of specific and observable behavior, how would you characterize your operation's current "safety culture?" If you don't have a handle on "the way things really are", where and when should you look to gain this insight?

2. What would you like your safety culture to be, described in specific terms of behavior?

Debrief Questions

1. Based on your data and first-hand observation, what are the key "targets for change" that you see as critical to creating the safety culture you want?

2. What are the key leverage points—people, places, events—you can take advantage of to speed up the change process and increase the likelihood that you will create the safety culture *you* want?

3. How will you measure success, and your progress in the process of change?

Chapter 15

Investing In Training

Fifty years ago, McGehee and Thayer summed up the potential return on investment in training, "The effectiveness of achieving an organization's goals will depend, in a significant way, on the nature and efficiency of the training employees receive for their assignments. " What's changed in fifty years? Only the breadth and depth of what employees need to know to be able do to meet those goals, starting with knowing what they need to know to be able to work safely. The need for understanding has never been greater.

Knowledge is no guarantee that anyone will work safely. But it falls to the leader to ensure that everyone knows what they need to know to be able to work safely. If they don't, it's not the training department's problem...it's your problem.

Put what your people know to the test. If you're not satisfied that they understand what they need to know, it's time to train. And, if training is what's needed, keep the Three T's in mind: timing, technique, and teacher. Properly combine all three, and the result can be memorable training--the kind of training that will help them, and you, achieve your goals.

And the ultimate goal of sending everyone home alive and well, every day.

Analysis Questions

1. What are the key things your employees need to know to be able to do their jobs safely?

2. How well does each of them measure up against what they need to know?

3. How well does each existing training program do in delivering the knowledge your employees need?

4. Since it's not possible to take on every challenge at the same time, what are the highest- priority courses in need of improvement in timing, technique and teaching?

5. What's the best path forward to improve training in these key need areas?

Chapter 16

Understanding What Went Wrong

When the consequences of a problem are big, a formal investigation is launched. Fortunately, they're rare, leaving all kinds of everyday, ordinary looking problems--often with little to no consequence—to the discretion of a supervisor. Often, the only difference between a major disaster and a non-event is found in the consequence. Yet, problems with little or no consequence can be the best opportunities to learn about what's really going on in your organization, and deal with problems before they become big.

At the core of every good investigation are the answers to the fundamental questions: who, what, when, where, how and why. The answers to all but the last question are found in the facts. So, find the facts first, and save the answers to the question why--conclusion, judgment, opinion--until after you've established the facts.

Understanding what went wrong is only half the battle; coming up with a solution that actually solves the problem is every bit as important. A good solution is one that meets two tests: 1) effectiveness, how well the solution fixes the problem; and 2) efficiency, how much time, effort and resources are required to implement the solution.

Finally, never forget that every investigation meeting meets the definition of a Moment of High Influence.

Preparation Questions

1. How often to you get reports of "minor problems" that don't reach the threshold of a formal investigation? What kind of problems do you hear about?

2. Of those problems, which ones do you think might represent the best opportunities to learn more about what's really going on in your organization, or to fix a problem before it produces significant consequences?

3. Which ones can you single out to better understand what went wrong?

4. What's your plan to take full advantage of one of these potential opportunities?

Debrief Questions

1. What problem did you choose to investigate?

2. How did you go about handing this Moment of High Influence?

3. What questions did you ask in the process of understanding what went wrong?

4. What specific questions did you find worked best to learn what went wrong?

5. The next time you have the opportunity to look into a problem, what will you do differently?

Chapter 17

Measuring Safety Performance

When it comes to safety, keeping score is easy: all you have to do is write down who got hurt. What's difficult is coming up with meaningful trend information in time to do something before the next injury happens. But that's exactly why coming up with better data the kind of data that can predict future performance is so important.

The solution is to apply the lessons learned about performance measurement from other applications: production, product quality, cost and even investing and competitive performance. A good safety performance measurement system fills two objectives: performance visibility—the ability to know what's really going on; and a competitive benchmark—the means of evaluating the results.

A simple but very powerful way to improve the measurement of safety performance is to measure both activities and results, and then sort them on when they show up in the performance cycle. Is it earlier or later? The combination of these four factors produces a Balanced Scorecard for safety. Finally, if you don't like the numbers you're currently counting, you can always create new and different ones.

Assessing Current Practice

1. What are the current safety performance measures your organization keeps?

2. How do they fit into the Balanced Scorecard?

3. Which ones are activities? Results? Which ones are Leading? Lagging?

4. How well do these measures do In providing you with performance visibility?

5. How well do these measures do in providing you with a competitive benchmark?

6. What's the current practice on reporting and measuring near-miss incidents?

Identifying Improvement Opportunities

1. What *don't* you know about the safety performance of those you supervise that you think you should?

2. What might you learn by "listening to the data" you already have available?

3. Given your assessment of the current state of safety performance measurement, what are the opportunities to improve how safety performance is measured?

Chapter 18

On The Horns: Managing Safety Dilemmas

A dilemma is a situation where two conditions exist, and are in opposition to each other. Managing safety performance is made even more difficult by the fact that leaders routinely find themselves on the horns of six significant dilemmas:

- **The Accountability Dilemma:** The gap between what a leader is accountable for, and what a leader actually can controls or influence.

- **The Risk Dilemma:** Operating with zero risk is impossible. Yet, when an injury occurs, the consequences are unacceptable.

- **The Investigation Dilemma:** Finding out *what* went wrong invariably leads to determining *who* did something wrong. But nobody wants to be found out to be that person.

- **The System Dilemma:** Individual human performance is a function of system performance. But, every individual is free to choose how he or she behaves.

- **The Middle Dilemma:** The middle level of management in the hierarchy plays the vital role of linking those at the top with those that do the work. Yet, leaders in the middle of any organization feel powerless, caught in the crossfire between those two levels.

- **The Leader Dilemma:** In producing great results, the most effective leaders are quiet and self-effacing. Yet, that is likely to leave the best leaders under-recognized and under-appreciated in the organizations they work for.

Managing these safety dilemmas is never easy. There's no getting around the fact that a dilemma pulls a leader in opposing directions, and it's all but impossible to eliminate either of their "horns." Dilemmas begin to become manageable when a leader first understands where the competing forces come from, and then develops a strategy for dealing with the dilemma.

Preparation Questions

1. In the past, when have you faced (or seen another leader face) these safety dilemmas:

   - The Accountability Dilemma

   - The Risk Dilemma

   - The Investigation Dilemma

   - The System Dilemma

- The Middle Dilemma

- The Leader Dilemma

2. What made dealing with the dilemma you faced difficult for you as a leader?

3. What dilemma are you most likely to face next?

4. The next time you face the dilemma, what will be your strategy to deal with it?

Debrief Questions

1. What specific dilemma did you face?

2. What were the competing forces that created the dilemma – for you?

3. What did you do to deal with the dilemma?

4. What worked well? What didn't go as well as you would have liked?

5. What have you learned that you can put to good use the next time this dilemma comes up?

Chapter 19

Leading From The Middle

As members of middle management, supervisors and managers play a vitally important role in the organization's hierarchy. Serving as the leader closest to those who do the real work of the organization gives a front line leader tremendous power that often goes unrecognized. So most front line leaders feel they have little influence over what goes on...and who goes home safe.

Yet no member of management has a better view of reality than does the front line leader. That knowledge is power. A front line leader has control—the ability to determine the outcome—over critical factors that have a major impact on safety performance: tools, methods, procedures, equipment condition, who gets assigned to do the work, their qualifications and competency. In the view of those doing the work, no level has more credibility and respect—and therefore more influence—than does the front line supervisor.

Leading from the middle of the organization is tough duty. The role comes with its own unique set of challenges, including the need to lead peers and bosses. But no level is more important to the goal of sending everyone home alive and well at the end of the day.

Preparation Questions

1. Why do you think many front line leaders fail to appreciate the organization power they actually possess?

2. One particularly important opportunity to "lead from the middle" is to provide upper management with a true picture of organization reality as it relates to safety. In your organization, where do you see a potential disconnect between what your leaders think is going on, and what is really going on?

3. What can you do to "upward lead" your leaders by providing a more realistic picture of organization performance?

4. What's your strategy to "upward lead…Who? When? How?

Debrief Questions

1. What was the opportunity you had to "upward lead?"

2. What did you do?

3. How well did it work out? What was the reaction by your leader?

4. Based on your experience, what will you do next time? What will you do differently?

Chapter 20

The Ten Biggest Mistakes Managers Make In Managing Safety Performance

When it comes to managing safety performance, managers have made more than their share of mistakes. These mistakes have often been made by well-intentioned leaders who fell victim to the popular management fad of the day.

The next generation of leaders has the opportunity to learn from those mistakes. The challenge of seeing to it that people work safely doesn't change much from one generation to the next. Understanding these mistakes what they were and why leaders made them—is an easy way to learn from someone else's errors, and to make certain you don't make them yourself.

The Ten Biggest Mistakes managers make in managing safety performance:

10. Thinking "Good Luck!" is Enough

9. Failing to Ask Good Questions

8. Driving Out All Fear

7. Focusing On The Short Run

6. Trying To Buy A Game

5. Dissing Safety Performance Measurement

4. Thinking Safety Management Doesn't Require Leadership

3, Trying To Manage Attitude

2. Believing Safety Performance Is Just Another Business Goal

1. Forgetting that Safety is Really All About Execution—and Failing to Recognize Who Manages Execution.

Discussion Questions

1. As a leader, what mistakes have you made?

2. What mistakes have you seen other leaders make?

3. Why do you think you—or they—made the mistakes?

4. In your role as a leader, where do you think you might be inclined to fall victim to the same thinking that produced the mistake?

5. What other "big mistakes" have you made managing safety performance—or witnessed other leaders make—that can represent learning opportunities for you and other leaders?

Chapter 21

Execution: The Name Of The Game….And Safety's Bottom Line

The challenges leaders face in sending people home safe—from compliance to complacency—boil down to **execution**. Safety execution involves getting everyone to faithfully follow all the rules exactly as intended and pay careful attention to what they are doing…all the time. For safety performance to improve, execution has to improve.

As simple as execution is, managing and improving it is one of the toughest challenges a leader faces. The Four Common—and Wrong—Assumptions help a leader understand why execution is normally far from flawless. The Four Absolute Truths About Execution begin to provide answers to what every leader needs to do to manage it.

Taken together, the leadership practices described throughout *Alive and Well* are all about execution. They provide the "what to do" and how to do it" to achieve the level of execution necessary to send everyone home safe.

Preparation Questions

1. How would you rate the current level of "safety execution" in your operation?

2. If you aren't certain as to what your operation's "safety execution" really is, how can you objectively find out?

3. For the areas and activities for which execution is not as good as you want it to be, what are the most important things to focus on improving?

4.   As the leader, what is your plan to improve execution in those key areas and activities?

5.   How will you measure improvement in execution?

Debrief Questions

1.   What are the specific areas and activities you selected to improve execution in?

2.   What techniques from Alive and Well did you utilize?

3.   What specific actions did you take as a leader?

4.   What worked well?

5. What could have gone better?

6. What improvements in execution have you seen?

7. What's next to further improve execution, or keep it at the level of excellence you have achieved?

Chapter 22

Can You Really Make A Difference?

When it comes to sending people home safe, leadership is the difference that makes the difference. There's no getting around that fundamental truth. Getting "common men to do uncommon things"—as Peter Drucker put it—is the stuff of great leadership.

Leading effectively in today's complex and demanding business environment represents a tough challenge. How well an organization's leaders collectively meet that challenge is reflected in the operation's injury rates.

The good news about safety leadership is that the practices that add up to making the big difference in performance are largely found in the everyday, ordinary activities of running the business. Don't be fooled by their ordinary appearance. These everyday-looking events are often significant Moments of High Influence, the best opportunity to lead.

Recognize them, lead well in these moments, and your operation will be a safer place for it. As a leader, there's nothing more important for you to do than that!

Final Questions

1. What have you learned about managing safety performance from reading and discussing *Alive and Well At The End Of The Day?*

2. What tools and concepts have you put into play that has helped lead the members of your crew to work more safely?

3. What will you do next to make a difference in the most important goal every business has... sending everyone home, alive and well, every day?

CPSIA information can be obtained at www.ICGtesting.com
Printed in the USA
BVOW05s0026310716

457337BV00006B/11/P